AF574363

HUDSON
PUBLIC LIBRARY
WITHDRAWN
FROM OUR
COLLECTION

ANIMALS OF THE CORAL REEF
Angelfish
by Kate Moening
BLASTOFF! 2 READERS
BELLWETHER MEDIA • MINNEAPOLIS, MN

**Blastoff! Readers** are carefully developed by literacy experts to build reading stamina and move students toward fluency by combining standards-based content with developmentally appropriate text.

**Level 1** provides the most support through repetition of high-frequency words, light text, predictable sentence patterns, and strong visual support.

**Level 2** offers early readers a bit more challenge through varied sentences, increased text load, and text-supportive special features.

**Level 3** advances early-fluent readers toward fluency through increased text load, less reliance on photos, advancing concepts, longer sentences, and more complex special features.

★ **Blastoff! Universe**

Reading Level

Grade K

Grades 1–3

Grade 4

This edition first published in 2022 by Bellwether Media, Inc.



Library of Congress Cataloging-in-Publication Data

Names: Moening, Kate, author.
Title: Angelfish / Kate Moening.
Description: Minneapolis, MN : Bellwether Media, [2022] | Series: Blastoff! readers : Animals of the coral reef | Includes bibliographical references and index. | Audience: Ages 5-8 | Audience: Grades 2-3 | Summary: "Relevant images match informative text in this introduction to angelfish. Intended for students in kindergarten through third grade" --Provided by publisher.
Identifiers: LCCN 2021000536 (print) | LCCN 2021000537 (ebook) | ISBN 9781644875025 (library binding) | ISBN 9781648344107 (ebook)
Subjects: LCSH: Marine angelfishes--Juvenile literature.
Classification: LCC QL638.P768 M64 2022 (print) | LCC QL638.P768 (ebook) | DDC 597/.72--dc23
LC record available at https://lccn.loc.gov/2021000536
LC ebook record available at https://lccn.loc.gov/2021000537



Editor: Elizabeth Neuenfeldt    Designer: Laura Sowers

Printed in the United States of America, North Mankato, MN.

# Table of Contents

# Life in the Coral Reef

queen angelfish

Angelfish live on coral reefs. They have **adapted** well to this **biome**.

Angelfish swim in warm ocean waters around the world.

## Queen Angelfish Range

Angelfish fins bend easily. This helps angelfish make quick turns.

Angelfish have flat bodies. They hide from **predators** in cracks between **corals**.

Angelfish can be almost any color.

This helps angelfish **camouflage** in colorful coral reefs. It also helps angelfish find **mates**.

blue ring angelfish mates

# Queen Angelfish Stats

| | | | | | | |
|---|---|---|---|---|---|---|
| Least Concern | Near Threatened | Vulnerable | Endangered | Critically Endangered | Extinct in the Wild | Extinct |

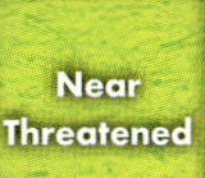

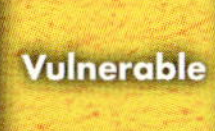

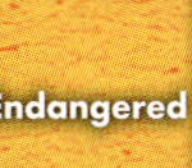

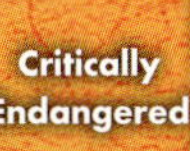

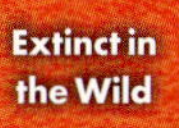

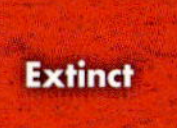

**conservation status: least concern**

**life span: up to 15 years**

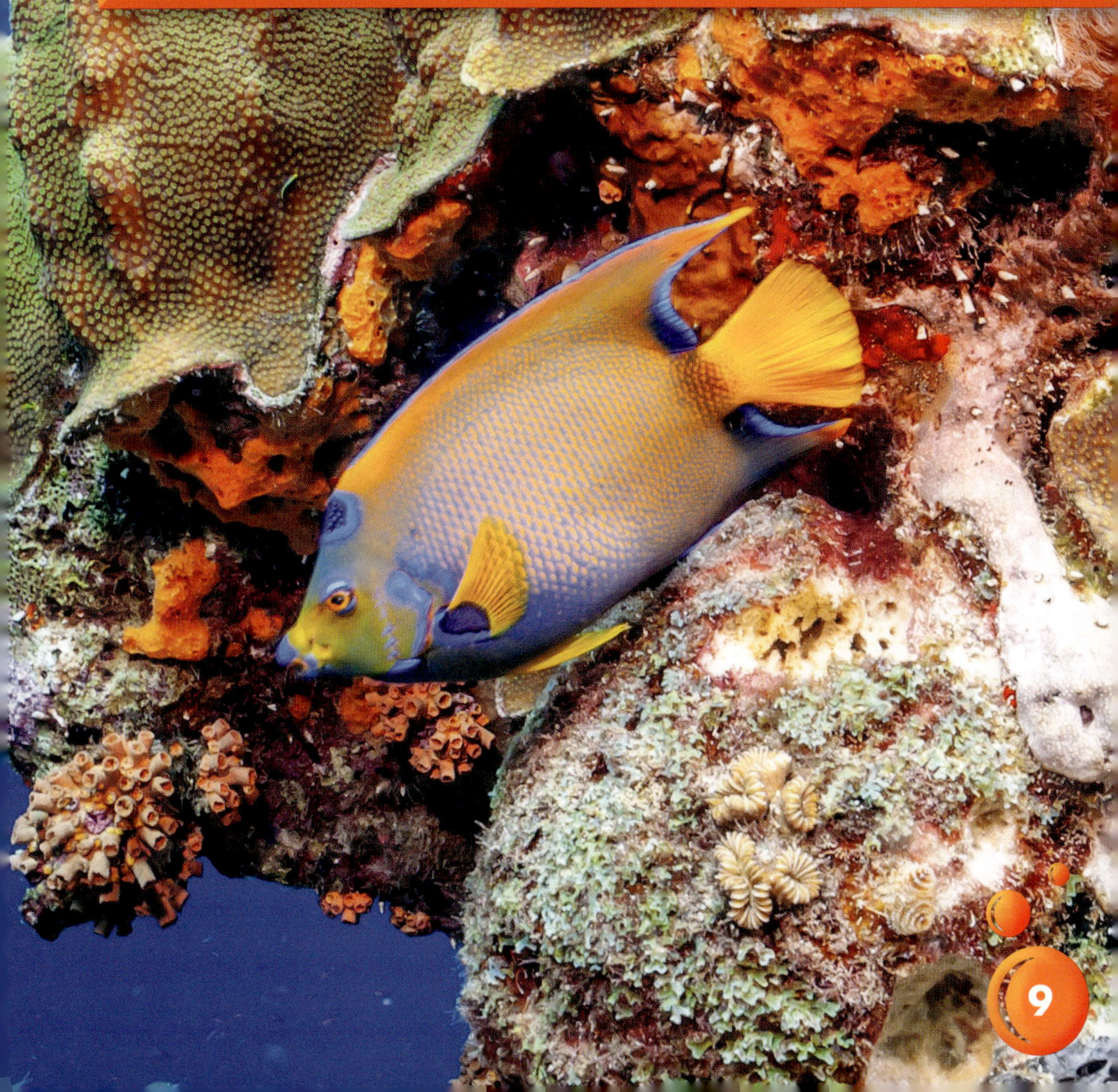

Angelfish have different markings. They can have stripes and spots.

flame angelfish

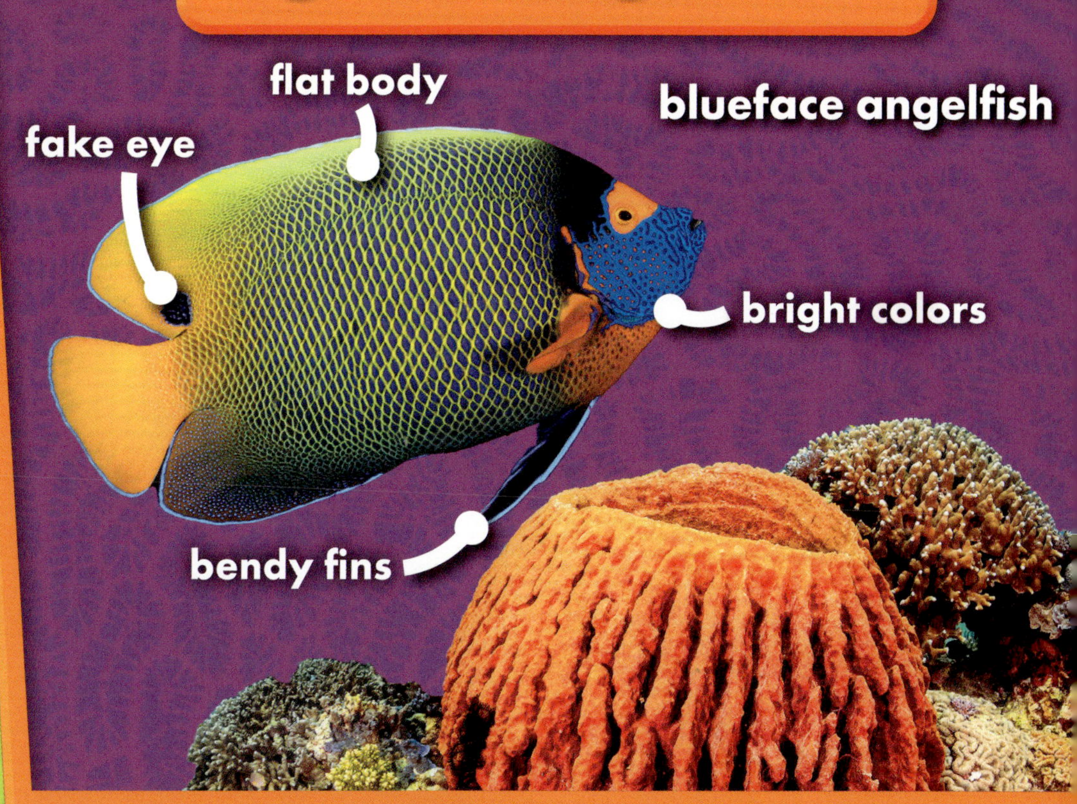


Some angelfish have spots that look like eyes. The fake eyes scare predators away!

# Hiding in the Reef

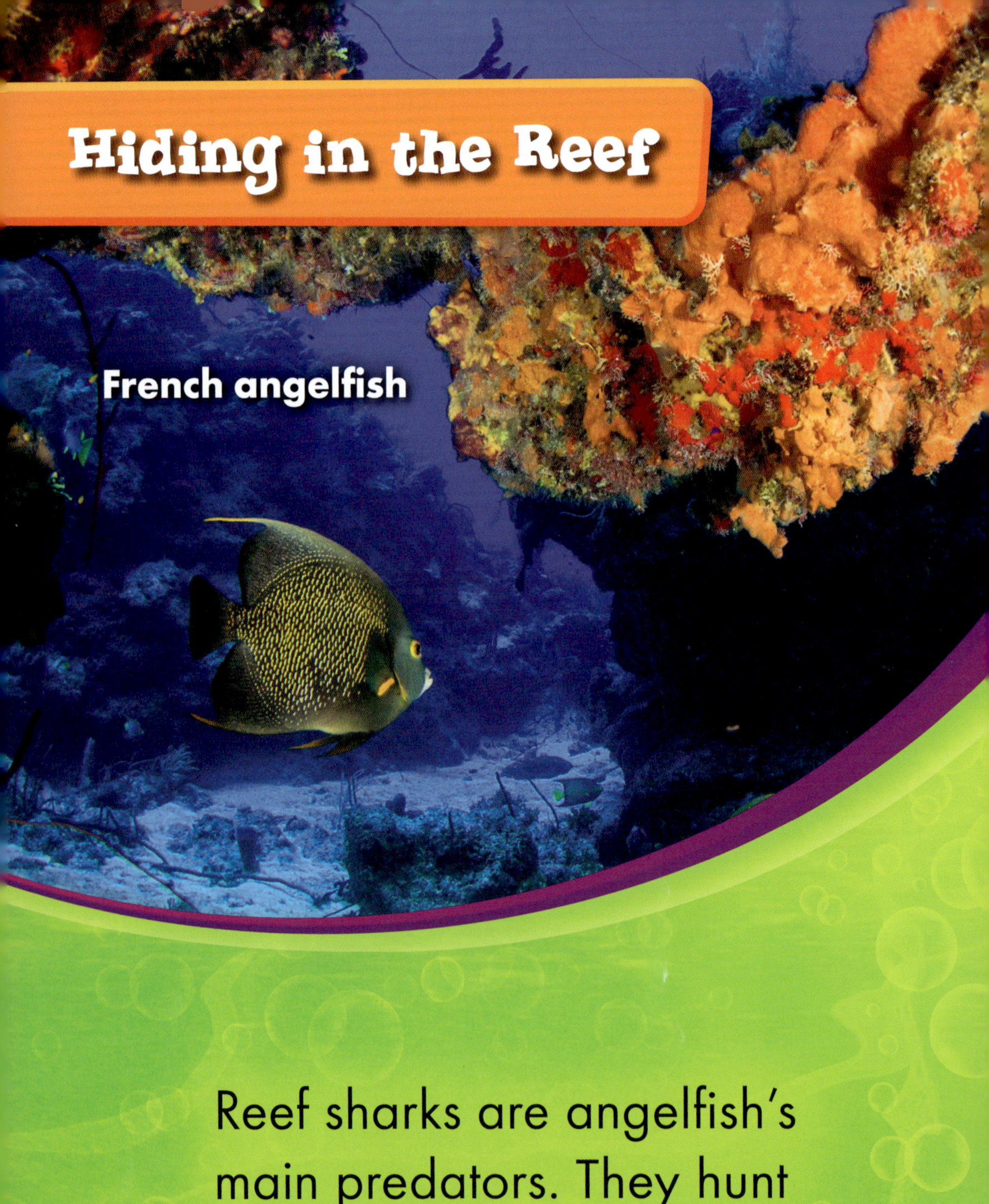

French angelfish

Reef sharks are angelfish's main predators. They hunt angelfish at night.

Angelfish hide from the sharks in the reef!

Some angelfish live in small groups. Groups have one male and several females.

Most angelfish live alone or in pairs. This makes it easier to find enough food.

# Reef Snacks

Cortez angelfish

Angelfish are **omnivores**.
Many eat **algae**.

Sponges are also a favorite food! Angelfish break through sponges with their small, pointed mouths.

Young angelfish often eat **parasites** off of bigger fish.

This helps the bigger fish stay clean. The young angelfish get a meal.

## Angelfish Diet

Angelfish look for food all day. They find hiding places at night to sleep.

Tomorrow will be another busy day on the coral reef. Sweet dreams, angelfish!

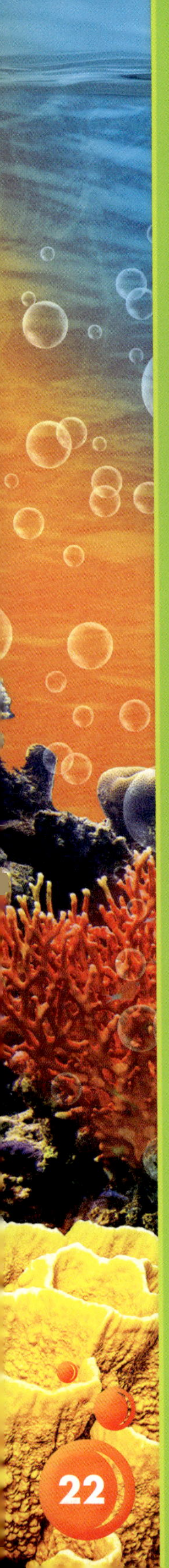

# Glossary

**adapted**—changed over a long period of time

**algae**—plants and plantlike living things; most kinds of algae grow in water.

**biome**—a large area with certain plants, animals, and weather

**camouflage**—to blend in with the surroundings

**corals**—the living ocean animals that build coral reefs

**mates**—partners

**omnivores**—animals that eat both plants and animals

**parasites**—living things that survive on or in other living things; parasites offer nothing in return for the food and protection they receive.

**predators**—animals that hunt other animals for food

# To Learn More

## AT THE LIBRARY


Keppeler, Jill. *20 Fun Facts About Marine Habitats.* New York, N.Y.: Gareth Stevens Publishing, 2022.

Rustad, Martha E.H. *Animals of the Great Barrier Reef.* North Mankato, Minn.: Pebble, 2022.

Shaffer, Lindsay. *Clownfish.* Minneapolis, Minn.: Bellwether Media, 2020.


## ON THE WEB

**FACTSURFER**

Factsurfer.com gives you a safe, fun way to find more information.

1. Go to www.factsurfer.com.
2. Enter "angelfish" into the search box and click .
3. Select your book cover to see a list of related content.

# Index

The images in this book are reproduced through the courtesy of: chonlasub woravichan, cover; Peter Leahy, pp. 4-5; RLS Photo, pp. 6-7; Lindsey Lu, p. 7; SergeUWPhoto, p. 8; John A. Anderson, pp. 8-9; PAUL ATKINSON, p. 10; Rich Carey, pp. 11 (angelfish), 22; Richard Whitcombe, p. 11 (coral); Drew McArthur, pp. 12-13; Matt Heath/ Alamy, p. 13; Damsea, pp. 14-15; blue-sea.cz, p. 15; Leonardo Gonzalez, p. 16; Brandon Cole Marine Photography/ Alamy, pp. 16-17; Nature Picture Library/ Alamy, pp. 18-19; Y-zo/ Wikipedia, p. 19 (Gnathiid isopods); Bill45, p. 19 (barrel sponge); Martin Habluetzel/ Alamy, p. 19 (green algae); George P Gross, p. 20; Mike Bauer, pp. 20-21.